SILKWORMS

by Liza Jacobs

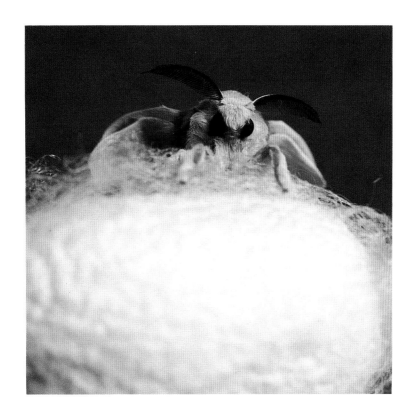

BLACKBIRCH® PRESS

San Diego • Detroit • New York • San Francisco • Cleveland • New Haven, Conn. • Waterville, Maine • London • Munich

© 2003 by Blackbirch Press™. Blackbirch Press™ is an imprint of The Gale Group, Inc., a division of Thomson Learning, Inc.

Blackbirch Press™ and Thomson Learning™ are trademarks used herein under license.

For more information, contact
The Gale Group, Inc.
27500 Drake Rd.
Farmington Hills, MI 48331-3535
Or you can visit our Internet site at http://www.gale.com

ALL RIGHTS RESERVED
No part of this work covered by the copyright hereon may be reproduced or used in any form or by any means—graphic, electronic, or mechanical, including photocopying, recording, taping, Web distribution or information storage retrieval systems—without the written permission of the copyright owner.

Every effort has been made to trace the owners of copyrighted material.

Photographs © 1992 by Chang Yi-Wen

Cover Photograph © Corbis

© 1992 by Chin-Chin Publications Ltd.

No. 274-1, Sec.1 Ho-Ping E. Rd., Taipei, Taiwan, R.O.C.
Tel: 886-2-2363-3486 Fax: 886-2-2363-6081

LIBRARY OF CONGRESS CATALOGING-IN-PUBLICATION DATA

Jacobs, Liza.
 Silkworms / by Liza Jacobs.
  v. cm. -- (Wild wild world)
 Includes bibliographical references.
 Contents: About silkworms -- A silky cocoon -- Feasting on leaves.
  ISBN 1-4103-0033-1 (hardback : alk. paper)
  1. Silkworms--Juvenile literature. [1. Silkworms.] I. Title. II. Series.

 SF542.5.J23 2003
 595.78--dc21
                                          2003001491

Printed in Taiwan
10 9 8 7 6 5 4 3 2 1

Table of Contents

About Silkworms 4

A Silkworm's Body 6

Shedding Skin 8

A Silky Cocoon 10

An Adult Comes Out .12

Soft and Hairy 14

Mating 16

Egg Laying 18

Feasting on Leaves 20

The Life Cycle 22

For More Information 24

Glossary 24

About Silkworms

4

Silkworms are not worms at all they are caterpillars! They spend most of their time eating. These white caterpillars mainly eat mulberry leaves.

Silkworms once lived in the wild, in Asia. But wild silkworms are not found anymore. More than 4,000 years ago, people in China began raising silkworms to collect the beautiful silk these caterpillars make. Today, people all over the world raise silkworms.

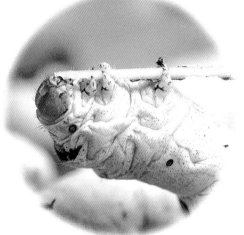

A Silkworm's Body

A silkworm's 3 pairs of legs are located just behind its head. Silkworms use their feet to grab onto the stems and leaves of the mulberry bush. They have strong jaws that help them eat their way through leaf after leaf. Near the back end of a silkworm is a small, pointed anal horn. Waste is passed through the nearby anal opening. Silkworms eat a lot, which means they also poop a lot!

Shedding Skin

In addition to their 3 pairs of legs, silkworms have 5 sets of fleshy pads that help them climb through the leaves. Silkworms eat constantly for about 6 weeks. The more a silkworm eats, the faster it grows. As a silkworm grows, it molts, or sheds its skin. The pictures running along the tops of these pages show a silkworm wriggling out of its skin. The old skin is left behind and a new, larger one is underneath it! Silkworms usually molt 4 times. Then they are ready to spin their silk cocoon.

A Silky Cocoon

Like other moth caterpillars, silkworms spin a cocoon.

But the single, beautiful, silky thread this caterpillar spins around and around itself is unlike any other.

Silkworms often get rid of the waste in their body before beginning their cocoon. Once it starts its cocoon, it takes 3-8 days for a silkworm to finish it.

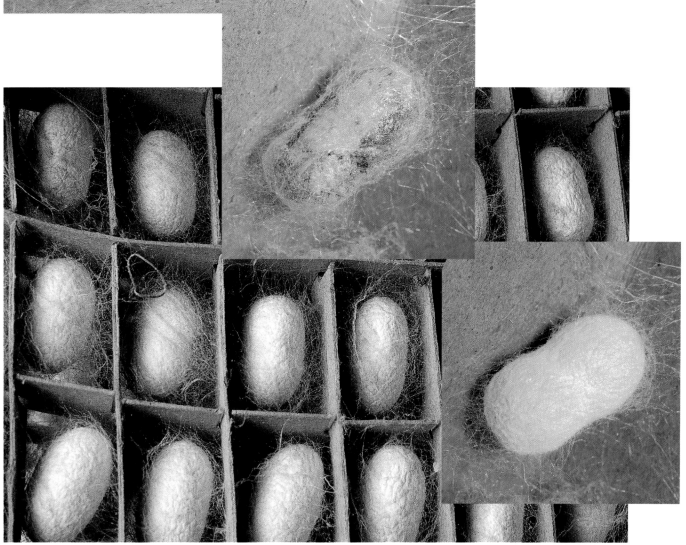

An Adult Comes Out

Inside its cocoon, the silkworm pupa is changing into an adult. The cocoon on these pages, which has been split open, shows some of what is going on inside. The silkworm's color darkens, it sheds its skin again, and begins to develop adult body parts. In about 3 weeks, the adult moth is ready to come out. Its saliva dissolves a hole at one end of the cocoon. Then the moth is able to push its body through the silk and come out.

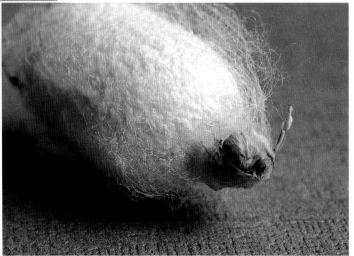
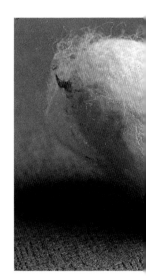

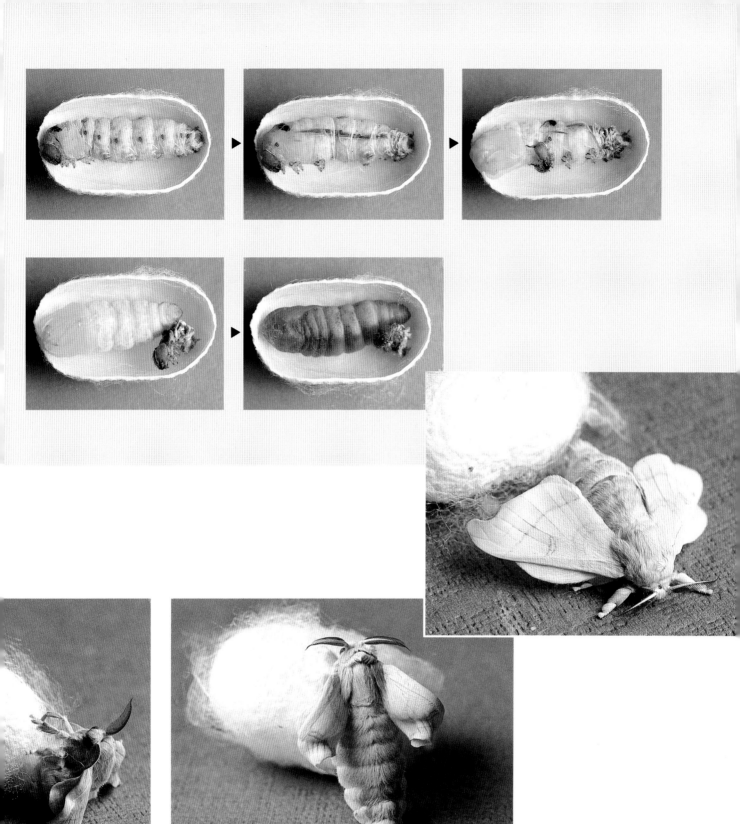

Soft and Hairy

Silkworm moths cannot fly. That is because their bodies are big and their wings are small. Their two pairs of wings and their bodies are ivory colored. Many have light tan markings. A silkworm s body is covered with soft hair. From its head, it has thick, comb-like antennae. Females are a bit larger than males.

Mating

Almost right after coming out of their cocoons, silkworm moths are ready to mate. In general, females do not move much. Males wriggle around in search of a mate, helped by their sense of smell. Once they find one, the mating process takes up to a day.

Egg Laying

After mating is complete, the female moth lays between 200 and 500 eggs. The tiny yellow eggs then darken and turn black. Silkworm moths do not eat. Both males and females die within about a week of coming out of their cocoons.

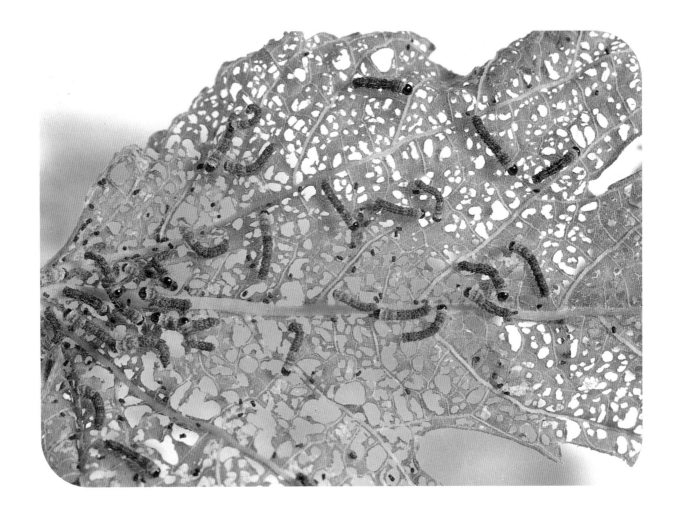

Feasting on Leaves

The eggs take about 10 days to hatch. Tiny silkworm babies, or larvae, push their way out of their shells. They are less than 1/4 of an inch long! For 4 to 6 weeks, the larvae do little more than eat, chomping hundreds of holes in the leaves. In fact, they eat so much that they increase their size by thousands of times! The dark larvae quickly grow into fat, white caterpillars that stretch about 3 inches in length.

The Life Cycle

Like many insects, silkworms go through an amazing cycle of life. A silkworm has four stages of life as it changes from egg, to larva, to pupa, and finally, to moth. With this unique insect, the pupa stage renders a cocoon that is a special and highly prized resource for humans.

For More Information

Drits, Dina. *Silkworm Moths*. Minneapolis, MN: Lerner, 2002.

Johnson, Sylvia A. *Silkworms*. Minneapolis, MN: Lerner, 1989.

Schaffer, Donna. *Silkworms*. Bridgestone Books, 1999.

Glossary

cocoon the protective case that moth larvae make to enclose their pupa stage

larva the second stage in a silkworm's life

molt to shed the outer skin or covering

pupa the third stage in a silkworm's life